Slithering Snakes

PYTHON

LONG AND STRONG

BY NATALIE HUMPHREY

Enslow PUBLISHING

DISCOVER!

Please visit our website, www.enslow.com. For a free color catalog of all our high-quality books, call toll free 1-800-398-2504 or fax 1-877-980-4454.

Library of Congress Cataloging-in-Publication Data

Names: Humphrey, Natalie, author.
Title: Python : long and strong / Natalie Humphrey.
Description: New York : Enslow Publishing, [2021] | Series: Slithering snakes | Includes index.
Identifiers: LCCN 2020005096 | ISBN 9781978517851 (library binding) | ISBN 9781978517837 (paperback) | ISBN 9781978517844 (6 Pack) | ISBN 9781978517868 (ebook)
Subjects: LCSH: Pythons—Juvenile literature.
Classification: LCC QL666.O67 H86 2021 | DDC 597.96/78—dc23
LC record available at https://lccn.loc.gov/2020005096

Published in 2021 by
Enslow Publishing
101 West 23rd Street, Suite #240
New York, NY 10011

Copyright © 2021 Enslow Publishing

Designer: Sarah Liddell
Editor: Natalie Humphrey

Photo credits: Cover, pp. 1 (python), 11 dwi putra stock/Shutterstock.com; background pattern used throughout Ksusha Dusmikeeva/Shutterstock.com; background texture used throughout Lukasz Szwaj/Shutterstock.com; p. 5 SarahB Photography/Moment/Getty Images; p. 7 Paul Starosta/Stone/Getty Images; p. 9 James Gerholdt/Photolibrary/Getty Images Plus/Getty Images; p. 13 Buddy Mays/Corbis/Getty Images; p. 15 amwu/iStock/Getty Images Plus/Getty Images; p. 17 Hillary Kladke/Moment/Getty Images; p. 19 Joe McDonald/The Image Bank/Getty Images; p. 21 okili77/Shutterstock.com.

Portions of this work were originally authored by Daisy Allyn and published as *Python*. All new material this edition authored by Natalie Humphrey.

All rights reserved. No part of this book may be reproduced in any form without permission in writing from the publisher, except by a reviewer.

Printed in the United States of America

Some of the images in this book illustrate individuals who are models. The depictions do not imply actual situations or events.

CPSIA compliance information: Batch #BS20ENS: For further information contact Enslow Publishing, New York, New York, at 1-800-398-2504.

Find us on f ○

CONTENTS

Big Snakes! 4
The Longest Snake 6
Python Babies. 8
Made for Hunting 12
Python Teeth 16
Sneaky Snake 18
Pythons and People. 20
Words to Know 22
For More Information 23
Index. 24

Boldface words appear in Words to Know.

BIG SNAKES

Found in Asia, Africa, and Australia, pythons are long giants! They live in many different **habitats**, but the places pythons like best are hot with plenty of trees for them to climb. Because they don't make **venom**, pythons hunt in a special way. They squeeze their **prey**!

THE LONGEST SNAKE

The longest snake in the world is the reticulated (rih-TIHK-yuh-lay-tuhd) python. Found in Southeast Asia, this giant snake can grow to be more than 30 feet (9.1 m) long. It can weigh up to 350 pounds (159 kg)! Reticulated pythons get their name from the **pattern** of their skin.

FEMALE RETICULATED PYTHONS ARE MUCH LARGER THAN MALES.

PYTHON BABIES

During the warmer months, female reticulated pythons usually lay between 25 and 50 eggs. If the weather is nice and there is plenty of food, they can lay over 100 eggs! The female will stay with the eggs to keep them safe until they **hatch**.

PYTHON MOTHERS KEEP THEIR EGGS WARM!

Baby reticulated pythons are only about 30 inches (76 cm) long. Since they are so small, they eat **rodents**, lizards, and frogs. Their pattern helps them stay safe by letting them blend into their homes to hide from **predators**. Hawks and other birds love to eat baby pythons!

BABY PYTHONS ARE READY TO HUNT AS SOON AS THEY HATCH.

MADE FOR HUNTING

Like many snakes, pythons use their tongue to help them find prey. Pythons also have special body parts called pits. A python's pits are found on its face, right behind its nose. Pits are used to feel the heat of passing prey.

Pythons aren't picky eaters and will eat anything they can fit in their mouth! Smaller pythons eat small animals, like frogs or lizards. Larger pythons eat larger animals, like monkeys or pigs. Some big pythons can even eat deer!

RETICULATED PYTHONS USUALLY EAT BIRDS AND RATS.

PYTHON TEETH

Pythons have sharp **fangs**, but they don't use them for killing. Pythons have four rows of fangs on the top of their mouth. These fangs all point **backward**. When a python bites its prey, the shape of its fangs helps keep the prey from getting away.

PYTHONS CAN HAVE UP TO 100 TEETH!

Sneaky Snake

Instead of chasing down their prey, pythons usually hide and wait. When prey walks by, the python bites it and holds on tight. Wrapping its long body around the animal, it squeezes hard enough to keep the prey from breathing. When the animal is dead, the python swallows it whole!

AFTER EATING, PYTHONS CAN GO MONTHS BEFORE THEY NEED TO EAT AGAIN!

Pythons and People

Pythons don't usually mind being around people. Some people keep pythons as pets or use them to kill pests in their home. Pythons are in more danger from people than people are from them! Some people hunt pythons and use their skin for belts, boots, and coats.

WHERE DO RETICULATED PYTHONS LIVE?

WORDS TO KNOW

backward Directed or turned toward the back.

fang A long, sharp tooth.

habitat The place or type of place where an animal naturally lives.

hatch To break out of an egg.

pattern A design that is repeated many times.

predator An animal that lives by killing and eating other animals.

prey An animal that is hunted or killed by another animal for food.

rodent A small animal, such as a mouse or rat, that has sharp front teeth.

venom Poison that is produced by an animal and is used to harm or kill other animals.

FOR MORE INFORMATION

BOOKS

Hamilton, S. L. *Pythons*. Minneapolis, MN: Abdo Publishing, 2019.

Orr, Tamra. *Reticulated Pythons*. Ann Arbor, MI: Cherry Lake Publishing, 2015.

WEBSITES

San Diego Zoo Kids
kids.sandiegozoo.org/animals/python
Learn more fun facts about pythons!

Wildlife Learning Center
wildlifelearningcenter.org/animals/asia/reticulated-python/
Check out more interesting facts about reticulated pythons!

Publisher's note to educators and parents: Our editors have carefully reviewed these websites to ensure that they are suitable for students. Many websites change frequently, however, and we cannot guarantee that a site's future contents will continue to meet our high standards of quality and educational value. Be advised that students should be closely supervised whenever they access the internet.

INDEX

baby pythons, 10
eggs, 8
fangs, 16
food/what they eat, 8, 10, 14
habitat/where they live, 4
hunting, 4, 12, 18
pattern of skin, 6, 10
pet pythons, 20
pits, 12
predators, 10
prey, 4, 12, 16, 18
size, 6, 10
squeezing prey, 4, 18
tongue, 12
trees, 4
venom, 4
weight, 6